AI 未來藝世界
繪圖新旅程

Arts Styles of the Future:
Non-Technical Introduction to AI Drawing

黃國禎　涂芸芳　金玲　楊梅伶　白璃　邱敏棋 —— 合著

推薦序

　　這本《Arts Styles of the Future: A Non-Technical Introduction to AI Drawing 未來藝世界：AI 繪圖新旅程》所表現的內容與價值令人讚嘆且感動。由生活中的人事物、童書、應用，乃至於文化與時尚，作者呈現的是融合人工智慧與藝術創作的極致構思與成品。尤其永續發展目標的主題，充分表現對地球的關懷，更讓這本書具有非凡的教育意義與參考價值。

　　作者之一的黃國禎教授是我多年的好友，他在教學與研究方面都有傲人的成就，更是國際知名的學者。很高興看到他與團隊成員合力把國際上最受重視的生成式人工智慧、藝術創作，以及永續發展目標等議題，作了最好的詮釋。在此極力推薦這本書給所有讀者。

臺中教育大學　校長

前教育部 資訊及科技教育司　司長

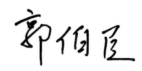

推薦序

　　我與黃國禎教授的交情已超過二十年，他一直以來都展現出優秀的研究能力、卓越的教學才華，以及不懈的實踐精神。這本書令我深感佩服，黃教授與他的團隊再度達到了全新的高度，將新興科技與藝術完美融合。

　　這本書所呈現的概念令人驚喜且讚嘆。透過人工智慧科技與藝術理論的結合，生活環境和生命的真善美得以展現。內容豐富多元，從多個角度深入探討了人工智慧應用於藝術創作的價值與魅力。對於那些渴望了解人工智慧繪圖與藝術創作的讀者來說，這本書無疑是目前最具參考價值的教材。

　　在此，我強烈推薦這本《Arts Styles of the Future: A Non-Technical Introduction to AI Drawing 未來藝世界：AI 繪圖新旅程》給所有讀者。這是一次難得的機會，讓我們一同探索人工智慧和藝術的奇妙世界。這本書將為您開啟一扇通往創意無限的大門，帶領您進入一場令人難忘的視覺盛宴。不要錯過這次獨特而富有啟發性的閱讀旅程！

臺灣師範大學　講座教授

中山大學　講座教授

推薦序

讓科技邁向美感與人文教育的新境界。

這本《Arts Styles of the Future: A Non-Technical Introduction to AI Drawing 未來藝世界：AI 繪圖新旅程》是協助準備初探 AI 繪圖夥伴們的極佳工具書。無論老師或是學生，都可以立刻有個概念，讓想像變真實，讓你我不受框架，能把創意發揮到極致。繪圖內容風格深具風趣、多元，從入門到娛樂到商業，讓人人都是小小詠唱師。如果你是沒有經驗的新手，本書可以帶給你無窮的樂趣，是一本令人讚嘆與感動的寶典。

作者之一的金玲老師是我非常敬佩與敬重的老師，人親切、和藹、溫暖，是位擁有斜槓人生的學者。老師除了是 AI ARTS 的研究者、教學者外，更是位心靈諮商師。她的這本書除了嘉惠師生外，對於新興科技相關教育的推廣上更具意義！

臺北市政府教育局 聘任督學
前臺北市立金華國民中學 校長

推薦序

學習新興科技，迎接與 AI 協作新時代。

目前 AIGC 產業前景極為看好，隨著人工智慧技術不斷進步，各個領域也紛紛帶來了巨大的機遇。許多企業也積極運用 AIGC 技術來實現創新。然而，在引領產業持續蓬勃發展的同時，我一直堅信，我們應首先推動教育領域的創新和進步，將創新的教學理念引入校園，使師生們都能夠受益。

而這本《Arts Styles of the Future: A Non-Technical Introduction to AI Drawing 未來藝世界：AI 繪圖新旅程》，不僅將藝術、科技和雙語教育融合在一起，更能快速引導任何人進入美感欣賞的藝世界。如果您有意涉足 AI 繪圖領域，這本書提供了多種題材和風格，絕對具有啟發性。

本書的其中一位作者金玲老師是我的後輩，她在國中、國小家長會事務上表現活躍，同時也積極參與將 AIGC 融入教育中。她不僅致力於培育種子教師，還常常透過研習演講等方式在校園內外推廣 AIGC。此外，在這本書中，她運用 AI 繪圖來呈現 SDGs 永續發展目標，讓我感受到情境與教學的結合，更深刻體驗到教育有著無限的可能性！

臺北市政府　市政顧問（教育組）

台北市國中、國小家長會聯合會　榮譽總會長

許坤仁

序

在《Arts Styles of the Future：A Non-Technical Introduction to AI drawing 未來藝世界：AI 繪圖新旅程》一書中，讀者將被引領進入 AI 與藝術主題的融合之旅。本書以淺顯易懂的語境，揭示生成式 AI (Generative AI) 在藝術領域的潛在能力，並引導讀者進入藝術風格探索的新世界。在本書中，所有的繪圖作品，都是透過知名的生成式 AI 繪圖工具 Midjourney 完成創作。

首章「Artistic Appetites: New Perspectives on Food through AI」，以多種視覺語言的風格，探討 AI 如何重新塑造對食物描繪的視角。接著，將讀者帶入「The Animal Kingdom: AI's Artistic Journey」的世界；在此章節中，從兒童繪畫到時光旅行風格，進而透過使用氣氛與情感的詞彙，以展示 AI 如何在動物描繪中展現獨特的視覺驚喜。

第三章「Future Faces: Exploring Portraiture and Creativity with AI」中，讀者將欣賞到表現主義、水彩速寫以及賽博龐克風格在肖像藝術中的應用，並進一步感受 AI 詮釋人物畫像藝術的無限可能性。在「Fashion Forward: The Evolution of Style with AI」一章，則運用 AI 素描技巧，展示時尚風格的轉變與進步。

在「Designing Spaces: The AI Approach to Interior Styles」章節中，讀者將透過建築與內部設計的視角，了解 AI 繪圖在室內風格創作上展現的魅力。「Elegant Transits: Showcasing the Beauty of Vehicles through AI Artistry」章節中，以機械風格、技術插圖，以及數位繪畫的形式，展現 AI 如何用藝術的方式呈現交通工具的美學。

在「Cityscapes: Urban Warmth Expressed in AI Art」章節中，透過超現實環境 (Hyper Realistic Environments) 描繪，讓讀者體驗 AI 所捕捉的城市溫度。「Celebrations in Canvas: Cultural Artistry in the AI Era」一章，介紹抽象表現主義與線條藝術如何與 AI 共同編織新的樂章。

「Storybook Magic: Unleashing Childhood Wonder through AI Illustrations」章節，結合 AI 運用各種插畫風格，喚起讀者內心的童真。「Sustainable Visions: The Importance and Promotion of SDGs in AI Art」章節，透過不同的風格描繪詮釋永續發展目標 (Sustainable Development Goals, SDGs)，以 AI 創作的角度來展現對地球的關懷。

最後，在「Artistry Unleashed: The Expansive Application of AI in Art」中，展示了 AI 在藝術領域應用的無限可能，包括彩繪玻璃、和紙膠帶、X-ray 雙重曝光、紙藝術與 ASCII 藝術等各式風格的創作，甚至還包含了 iPhone UI/UX、貼紙、富有表現力的漫畫面板等多元面向設計的實踐。

期待讀者能透過這本書，探索 AI 與藝術的美麗交融；相信無論是藝術愛好者還是科技愛好者，都能在其中找到共鳴與新的啟發。讓我們一起透過這本書，進一步了解 AI 與藝術交織的多彩世界，體驗到藝術不再是遙不可及的美麗。

"AI 藝術帶來的悸動如彩雲，讓人不禁著迷，願乘風追隨。　黃國禎"

"AI 藝術如同奇異的夢境，繽紛燦爛，引領我們踏上充滿驚喜的創意之旅。　　　　　　　　　　　　　　　　　　涂芸芳"

"一點創意由心起，森羅萬象為君顯。　　　　　　　　金玲"

"藝術是魂魄的倩影，以無言的語言觸動你我的靈魂。　楊梅伶"

"AI 與藝術相遇，點亮無盡創意。　　　　　　　　　白璃"

"AI 為藝術與科學的跨界發展帶來無限想像空間。　　邱敏棋"

探索本書的附加資源
（其中包括如何製作
LINE 貼圖的完整流程）

Contents

1 Artistic Appetites: New Perspectives on Food through AI

2 The Animal Kingdom: AI's Artistic Journey

3 Future Faces: Exploring Portraiture and Creativity with AI

4 Fashion Forward: The Evolution of Style with AI

5 Designing Spaces: The AI Approach to Interior Styles

6 Elegant Transits: Showcasing the Beauty of Vehicles through AI Artistry

7 Cityscapes: Urban Warmth Expressed in AI Art

An image showing many plates of food on a white dinner table, in the style of cloisonnism, elaborate fruit arrangements, he jiaying, sultan mohammed, frequent use of yellow, affandi, skillful

Artistic Appetites: New Perspectives on Food through AI

Contemporary Realism

當代寫實主義 以奇特的情境與物體組合挑戰超越現實的創新視覺效果。具有開放性、多樣性和挑戰性的創新特點。它不只顛覆傳統現實，更探索人類內心深處，讓觀者感受想像力與創新的力量。

Healthy Recipes

Healthy recipes to celebrate christmas with family and friends, in the style of pigeoncore, red and amber, iberê camargo, traditional, Contemporary Realism

Candied Oranges Dessert

Dessert Topped with Berries

Asian Dessert on a White Plate

Cartoon Style

卡通風格 以鮮豔的色彩和誇張的形象為主要特點。該風格通常透過擬人化或具有視覺張力的方式來表達故事和情感，經常使用超自然的色彩來強化視覺吸引力。

Stinky tofu

Delve into the world of Cartoon Style Drawing with a lively depiction of Stinky Tofu. Let the bold lines and vibrant colors bring this unique dish to life. Capture the distinct texture and aroma with playful expressions and exaggerated features. Imagine the sizzle and steam as the Stinky Tofu is cooked to perfection. Embrace the cultural significance and culinary adventure in this dynamic Cartoon Style Drawing representation

Dim Sum

Hotpot

"簡單樸實的地方美食，融合了文化、創意，以及烹調者的巧思。"

Fried Dumplings

Chinese Spring Rolls

Watercolor Painting

水彩畫風格 以透明度和自然流動性為顯著特點。透過色彩層次與深淺變化，該風格用簡潔而細膩的筆觸來傳達情感和氣氛。

Dongpo Pork

Drawing picture of Dongpo Pork,grilled very thin slice beef wagyu, cilantro on top, for advertisement in cartoon, pastel color, with white background, Watercolor Painting Style

Steamed Fish in White Wine Sauce

Tonkotsu Ramen

Korean Fried Chicken

Beef Noodles

"精湛的廚藝，讓平凡的食材展現不平凡的美感與口感。"

Roasted Lamb Chop

Cheeseburger

Pizza & Red Wine

Shrimp

▼ Wooden Board Bread & Olive Oil

"無論在那個國家，來自什麼文化，色香味俱全的美食所帶
來的歡樂，永遠是生命中最美好的回憶。"

▲ French Cuisine

Cake Ingredients & Treats

line sketch, Infographic, watercolour, cake ingredient and treats, cartoon sketch style

Different Kinds of Cakes Drawn on the Chalkboard

Line sketch, illustration for children, many different kinds of cakes drawn on chalkboard, in the style of dmitry spiros, pop art aesthetic, stephen shortridge, sui ishida, y2k aesthetic, ferrania p30, emotive fields of color

12 Zodiac Animal signs, Neo-Dada Pop Art

The Animal Kingdom:
AI's Artistic Journey

Children's Drawing

兒童繪畫風格 以其純真與創意著稱。該風格
呈現了孩子們的無邊想像力與自由表現的本能。
從簡單的線條圖案到豐富多彩的插畫，兒童的繪
畫充滿了歡樂與無窮的可能性。

Yellow Mouse

A children's drawing of a yellow mouse, poor quality

Yellow Monkey

Yellow Dog

Yellow Horse

Yellow Bird

"童畫不只展現童真，更勾起無限的回憶；看到這些簡單且趣味的圖面，回想到孩童時期最幸福無憂的日子。"

▲ Dragon (Colored Crayons, Bright Vibrant Colors)

Mouse

Fluffy Bee Baby

Cat

Rabbit

A black and white drawing of a cute {mouse, fluffy bee baby, cat, rabbit} in the middle of a storm, in the style of gond art, gigantic scale, bamileke art, furry art, woodcut, full body, inuit art

Art Brut

原始藝術風格 以獨特、非典型和直觀的方式
聞名。此種藝術形式打破了傳統的美學規範，透
露出藝術家內心最真實的情感。它彰顯了個人創
造力與自由表達的精神。

Cute Monkey

A cute monkey, art brut

A {cute sheep, rabbit, French bulldog dog, Persian cat} wearing reading glasses, a beanie and a black and white collar, artwork by Steve Dillon and Shusei Nagaoka

Hyper realistic photograph of a {penguin, beautiful rooster, common tortoise, white snake} on black background, studio photo, sidelight

Impressionism

印象主義 以捕獲瞬間光線和感受的獨特風格而著稱。該風格強調色彩和光影的流動變化，透過快速且混合的筆觸，呈現出模糊卻又生動的視覺效果。印象主義作品追求對自然光線與情緒氛圍的精確捕捉，展現出對自然現象和觀察力的敏銳洞察。

Ox

A Ox of Impressionism

Time Travel

時光旅行 讓我們穿梭於不同的時代，每個時期都具有其獨特的視覺風格和藝術魅力。從古典到現代，時代的不同展現了各自的視覺美學。

`/imagine prompt <decade> dog illustration`

Impasto is a Dog of the Library (Impasto Painting)

Fauvism Style

野獸派藝術風格 以大膽的鮮豔色彩而聞名，它強調情感的表達和形式的簡化。該風格特點是使用活潑的筆觸和非傳統色彩，對現代藝術產生了深遠影響。

Cute Cat

A cute cat of fauvism style

Cute Tiger

A cute tiger of fauvism style

Emote

讓 AI 藝術與 **情感** 詞彙的融合，不僅為角色賦予了豐富的個性，還增強了讀者的情感共鳴。情感詞彙涵蓋了各種情緒，使故事變得更加多元與引人入勝。AI 藝術開創了創作新的可能性，並進一步豐富了故事的深度和感染力。

/imagine prompt <emotion> Koala

Happy Koala   Determined Koala   Shy Koala

Angry Koala   Sleep Koala   Embarrassed Koala

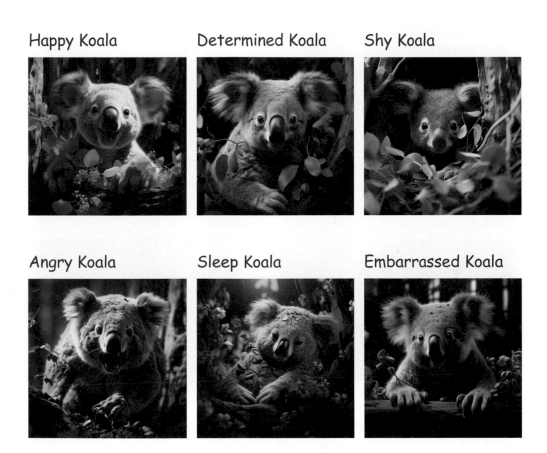

Composition Perspective/Perspective in Composition

視角 選擇在攝影和繪畫中是一項重要元素,對於視覺效果和主題的表現有著關鍵影響。以下是一些常見的*視角*選擇:

- 正視角 (Front View):直接正對被拍攝或繪畫的對象。
- 後視角 (Back View Angle):從對象的背面觀察或描繪。
- 側視角 (Side View or Shoulder Level View):從側面或相當於肩膀的高度觀察或描繪對象。
- 平視角 (Eye Level View):在眼睛水平線上的視角,這是最常見的視角。
- 微觀視角 (Microscope View):放大觀察細節的視角,像是在使用顯微鏡。
- 廣角視角 (Wide Angle View or Super Side Angle View):以更寬闊的視角來觀察或描繪對象。
- 鳥瞰視角 (Bird's Eye View or Top-Down View):從上方看下去的視角,猶如鳥兒在空中俯瞰。
- 蟲視、仰視 (Worm's Eye View or Bottom-Up View):像是從下方看上去的視角,譬如從地底看向天空。
- 魚眼視角或魚眼透視 (Fish Eye Perspective):一種特殊的廣角鏡頭視角。這種視角會將視線中的景象強烈地扭曲和放大,以達到一種全景的效果。
- 遠景視角 (Long Shot View):從遠處看對象,將環境納入視角中。

Wide Angle

Microscope View

- 近景 / 特寫視角 (Close Up View)：近距離放大對象的視角，注重對象的細節。
- 全景視角 (Panoramic View)：將廣闊的景象完全納入視線的視角。

/imagine prompt A adorable cat in the garden, <Composition Perspective>

Back View Angle

Side View

Bird's-eye View

Fish Eye Perspective

Expressionism, Smiling children holding hands around the world, Children's Day concept, clear outline, high detail

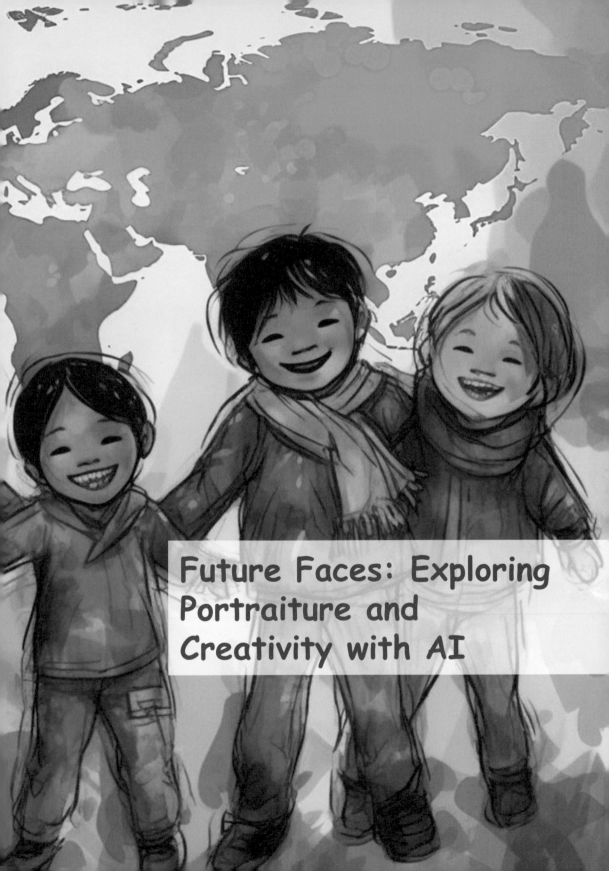

Future Faces: Exploring
Portraiture and
Creativity with AI

Expressionism

表現主義風格 透過濃烈的主觀情感來描繪以展現內心的深處和複雜性。此風格不僅僅著重於物體的實際呈現，更是藉由藝術手法的扭曲和變形，喚起觀者的情感反應和思考，將焦點集中在主觀感受和內在情感上，以此揭露出人類的情感與思想。

Beautiful Goddess of the Seasons

Expressionism, the beautiful goddess of four seasons

The Back of Cute Elementary School Students Going to School

▼ Grandpa & Grandma

　　"最刻骨銘心的情感莫過於透過人物的表情及肢體語言
帶來的衝擊；無論是快樂、悲傷，或是思念，這樣的衝擊
是那麼令人心動。"

▲ Little Boy & Beautiful Woman Holding Hands

A Girl in a Hat and a Dog Running in the Sunshine

Watercolor Sketch

水彩速寫風格 以自然透明與生動特性引人入勝。以簡單線條與色塊描繪物體，水彩的流動性和透明度營造深淺變化，使作品更具生命力。它不只呈現實體形態，更透過色彩和筆觸變化傳達藝術家情感與視角，讓觀者感受內心的細膩與豐富。

A Girl Holding a White and Yellow Tiger

Fixed camera, watercolor sketch, simple Clip art, An oil painting of a girl holding a white and yellow tiger, in the style of anne stokes, josephine wall, chiaroscuro portraitures, red, animated gifs, vladimir volegov, portrait, fujifilm natura 1600 realistic line art

A Lady Laying on a Plush Sofa, Sharing Tranquility with a
Gorgeous Cat

A Woman in a White Dress Leads a Tiger Through the Forest

Cyberpunk Style

賽博朋克風格 以深邃的科技視覺和未來主義叛逆元素描繪出高科技且道德混亂的社會。透過鮮豔霓虹與陰鬱城市景象的對比，揭露對未來的觀察與想像。它不僅再現了現實，更深入探討了社會矛盾與人類的內心世界。

Artificial Intelligence Robot in the Cyberpunk Style

Cool Motorcycle Handsome Man

Korean version, Cool motorcycle handsome man, helmet in hand, medium and long hair micro curly, face close - up photo, background road, riding on a motorcycle looking at the camera, cyberpunk style.

Robot and Machine Princess Double Exposure Photography, Double Exposure Photography, Cyberpunk

▼ Female Robot Futurism, Cryptopunk

"超出現實的視覺效果，彷彿在訴說未來人類與科技
密不可分的關係。"

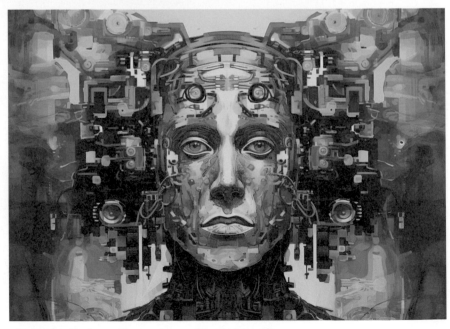

▲ Male Robot Futurism, Cryptopunk

The Jungle Goddess

Two anime faces with eye masks on each side, in the style of surrealistic horror, intricate illustrations, intricate black and white illustrations, stains/ washes, puzzle-like pieces, striped painting, made of all of the above --niji 5

URL Simple outline of cartoon tomamto, minimalist, white background, line art, 2d flat file, clipart

URL Minimalist, black and white logo, in the style of Paul Rand, a cute boy

URL 3D Cute kawaii boy, geometric 3D style

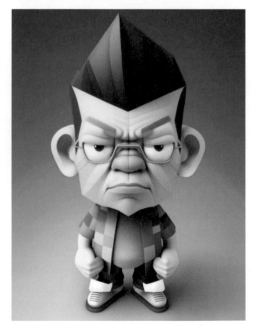

URL Cute sweet kawaii boy, geometric 3D style

#URL 在這個情境中被用來指示一張基底圖片的位置。

Portrait

肖像畫風格 能精細捕捉人物的面部特徵和內在情感，透過色彩、線條和光線的巧妙運用。它不只再現物理外貌，更描繪個體獨特的精神世界，為觀者與被描繪者建立情感連接，呈現人性的多樣性。

Male Teacher

URL An animation crossing cute cartoon male teacher immersed in teaching and research, by John Smith, academic portrait artist, a scholarly atmosphere, casual clothes, intense concentration, a chalkboard filled with complex equations, piles of books and papers, an open laptop with data analysis, walls adorned with educational posters, rich wood desk, inspirational quotes, a passionate figure of knowledge without spectacles, making difficult concepts easy

#URL 在這個情境中被用來指示一張基底圖片的位置。

Handsome Man

Medium

生成時尚圖像的最佳方法之一：指定一種**藝術媒介**。

/imagine prompt <any art style>, portrait, a cute cartoon young man

Pencil sketch style

ink painting

Watercolor sketch style

Oil painting

Digital painting

Block print style

Cyanotype style

Risograph style

Ukiyo-e style

Paint-by-Numbers style

Abstract painting

Expressionism

Graffiti style

Folk art style

Pop art style

Customized Line Stickers - 帥男神貼圖

AI 繪圖在 Line 貼圖的應用非常廣泛。可以使用 AI 繪圖技術創建自己的個人化貼圖，以反映你的風格、興趣和個性。我們可以使用 AI 繪圖的模型將照片 (圖片) 進行不同風格處理。

A set of hand-drawn sketches of Early Summer Women's Clothing, Pencil draw, Design drawings, Fashion photography, Colorful, Clothing, HD 8K --v 4

Fashion Forward: The Evolution of Style with AI

Sketch

時尚設計的 *素描風格* 以流暢的線條展現服裝的輪廓和細節，強調流線和穿著效果。藉由素描，設計師能快速捕捉創意、傳達服裝的風格和質感。這些素描作品是設計過程中的重要工具，也展示時尚設計師的才華。

Summer Men's Clothing

A set of hand-drawn sketches of Early Summer men's Clothing, Pencil draw, Design drawings, Fashion photography, Colorful, Clothing, HD 8K

Modern Fashion Sketch,
Capsule Collection of
Women's Clothing

Gorgeous Chinese Hanfu Wedding Cloak

Evening Gowns

"典雅飄逸的美感，融合知性與感性的丰采，
吸引眾人的目光。"

Evening gown croquis, inspired by elegant fairies and banana-split-ice cream, beautiful silk crepe and glittering sequins, layered, tea length, strapless, fairytale princess core

Early Summer Women's Clothing

Early Summer Men's Clothing

"美麗、野性與帥氣，令人無法抗拒，唯有摒息注目。"

A supermodel walks the catwalk in a pink shiny super-fat puffer jacket, inflatable, transparent, houte couture, shiny, glass, high res

A male supermodel walks the catwalk in a deep blue shiny super-fat puffer jacket, inflatable, transparent, houte couture, shiny, glass, high res

An artful portrayal of a De Stijl academic library, modern architectural illustrator, strict lines, bold use of primary colors, an architectural study in abstraction and simplicity, encapsulating the universality of learning in its unique design, a distinctive interpretation of knowledge dissemination

Designing Spaces:
The AI Approach to
Interior Styles

De Stijl Style

荷蘭風格 以簡潔、幾何形狀和基本色調 (紅、藍、黃) 與黑灰白色調聞名。
透過精心的結構和色彩平衡,營造出淨化、整齊以及和諧的視覺效果。這種風格
強調整體秩序,淡化個體差異,傳達普遍的視覺語言。

Floor Plan

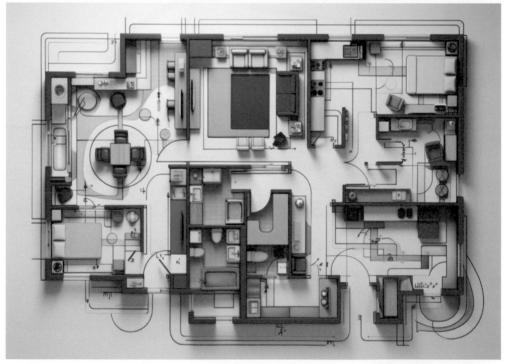

De Stijl style, floor plans for a two bedroom apartment with a bedroom, living
room and bathroom, in the style of with red, yellow and blue, simplified line
work, northwest school, computer-aided manufacturing, contour line, AGFA
vista.

Interior Design of a Living Room

Modern Warm Bed with Geometric Pattern Wall Art

Mondrian Painting in Isometric Style

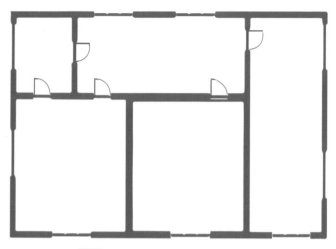

 以圖為基底 (URL) 進行繪製

Flat Illustration
One Storage Home Floor Plan, 2 Bedrooms

Interior Design

室內設計 注重空間與美感的平衡，融合功能性與舒適性。透過顏色、燈光、家具與材質的巧妙搭配，創造出反映個人風格的和諧空間，並深刻影響我們的情緒與生活品質。其包含 Modern Style, Industrial Style, Scandinavian Style, Country Style, Mediterranean Style, Classic Style, Luxury Style, American Style, Japanese Style, Chinese Style, LOFT style, Victorian Style, and Minimalism Style 等。

LOFT Style

URL perfect interior design, LOFT style, home furniture, carpet, very realistic
#URL 在這個情境中被用來指示一張基底圖片的位置。

Minimalism Style

Modern Style

Bohemian Style, Interior Design of a Living Room

Bauhaus Style, Dark Yellow, Interior Design of a Living Room

▼ Rococo Style, Interior Design of a Living Room

"典雅舒適的生活品味，猶如在夢想中的國度，
令人陶醉。"

▲ Scandinavian Style, Interior Design of a Living Room

▼ Line art, Watercolor, Modern Office Interior in Loft, Industrial Style

▲ Conceptual Sketch and Architectural Plans

▼ Sketch, Bathroom Showing a Bathtub

"融合自然景觀的居家設計，讓人不禁讚嘆：世界上再也沒有比住在這裡更幸福的事了！。"

▲ CSLC Campus Plan and Design, in the Style of Meticulous Sketches, Modern European Ink Painting

Sketch Outline of the Wooden Table in the Library Rouge Chair and Book

Sketch Style Drawing of a Modern Style Library Design

De Stijl Academic Library, Modern Architectural Illustrator

Architecture Interior Living Room, 3D Sketch Drawing

A Stunning Translucent Parametric Architectural Structural Installation

▼ Modern Home Interiors with Stained Glass Floor, Industrial Style

"充滿未來感的生活空間，有穿越到另一個時空的感覺。"

▲ Modern Home Interiors with Stained Glass Floor, Industrial Style, Pink light

Create 4D image of F1 car with a driver in F1 race in a well-known circuit with sponsor's logo on its body, in the style of vibrant, time lapse, technical illustration

Elegant Transits: Showcasing the Beauty of Vehicles through AI Artistry

Mechanical Style

機械風格 以精密的細節和結構
複雜性為特點。該風格傳達出科
技與工程的力量,透過刻畫精確
的機械元素與組件,展現其功能
與美學。機械風格不只描繪物理
實體,更透過其細節,讓觀者領
略機械之美,並體驗創新與工程
的力量。

Steampunk Steam Engine Train

Steampunk steam engine train, mechanical style

▼ Bus

Hot Air Balloon

Helicopter

"古人的工藝與智慧，融合了技術
與古典的美感，令人愛不釋手。"

▲ Carriage

Gyroscopic Weird Car

Food Truck

Dune Buggy

Mass Rapid Transit

Technical Illustration

技術插圖 是一種以精確度和詳細性為
特色,用以傳達技術資訊或概念的插畫方
式。該風格強調將複雜的技術性信息轉化
為直觀且易於理解的視覺表示。透過精確
的線條和清晰的形狀,技術插畫不僅詳細
呈現出物件的形態與功能,同時專注於細
節和準確性,讓讀者能更了解其運作原理
和結構設計。

Canoe

Canoe, Technical Illustration

Cable Car

Dodge Brothers Touring Car

Human Rickshaws

Digital Drawing

數位繪畫 是一種結合電腦、繪圖板和專用軟體 / 數位工具的創作方式。其能夠模擬各種傳統繪畫手法或創新的視覺效果，進而提供更大的創作自由度，以呈現內心世界和豐富的想像力。這種藝術形式代表了藝術與科技的融合，並展現了其無限的創新可能性。

Fishing Boat

Fishing boat,watercolor with pen outline, pastel color, white background, digital drawing

Ambulance

Bicycle

Motorcycle

"嚮往自由與無拘的
生活型態,以及探索
這個世界的快樂。"

Vintage Airplane (Mechanical Style, Digital Drawing)

Ford Model T (Vintage Style)

Vintage style, hand drawn, includes a three - dimensional view of a Ford Model T, industrial design, colorful, white background

Airplane (Children's Style Drawing)

Formula One Car (Children's Style Drawing)

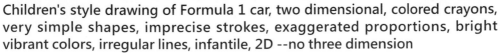

Children's style drawing of Formula 1 car, two dimensional, colored crayons, very simple shapes, imprecise strokes, exaggerated proportions, bright vibrant colors, irregular lines, infantile, 2D --no three dimension

Inkblots hyper realistic environment painting of an old busy street under Mt Fuji Japan, realistic, diagrammatic, impasto paint, nostalgic yet abstract, historical scenes sketch

Cityscapes: Urban
Warmth Expressed in
AI Art

Hyper Realistic Environments

超現實環境風格 為了打造充滿意境的藝術世界，將超現實概念帶入到各國知名的街景中，創造出意境豐富的藝術世界。透過想像讓我們可以倘佯在藝術化的世界。從平面的文字轉為充滿意境的類現實，並穿插於不同年代，讓共識現實能直接感知。

Soeul, Korea

Old streets of Soeul Korea, hyper realistic environments, hard-edge painting, post processing, historical scenes sketch, 1960s Korea, rural town center, town square, brick buildings

Taiwanese Children Playing on the Roads

Ho Chi Minh, Vietnam

Kowloon, Hong Kong

Taimali Station, Taiwan

Rome, Italy

Seville, Spain

▼ Sydney, Austrilia

"最美的街景，夢想的空間，源自生活與藝術的結合，
以及人們的審美觀。"

▲ Paris, France

Santorini, Greece

Jungfrauioch, Switzerland

▼ North Island, New Zealand

"當生活的空間與大自然融合，什麼是現實，什麼是夢境，
已經不重要了。"

▲ Cappadocia, Turkey

Giza Pyramids, Egypt

Chiang Mai, Thailand

▼ Amsterdam, Netherlands

"每一個生活的縮影，都代表了千百年的文化與習俗。"

▲ Tokyo, Japan

London, United Kingdom

Lisbon, Portugal

"繪圖展現了真實的生活，也展現了我們的響往。"

▲ Quebec, Canada

Las Vegas, United States

Casablanca, Morocco

▼ Patagonia, Argentina

"繪圖不只展現文化，也展現生活的趣味及歷史的足跡。"

▲ Little India, Singapore

Light watercolor, Winter fairytale, bright, Merry Christmas greeting card with copy-space, Happy snowman standing in Christmas landscape, Snow background, few details, dreamy Studio Ghibli

Celebrations in Canvas: Cultural Artistry in the AI Era

Abstract Expressionism

抽象表現主義 是一種以抽象形式和情感表達為主的藝術風格。它強調藝術家對內在情感和精神世界的表達，將內心深處的情感和思想轉化為獨特的藝術作品。該風格通常以大膽的筆觸、濃厚的顏料和抽象的形式呈現，追求對生命和存在的深度探索。

Valentine's Day

Valentine's Day, scent, a couple, science, bio, monotone style, Abstract expressionism style

Teacher's Day

Father's Day

Happy Chinese New Year (watercolor+ink+gold)

Lantern Festival (watercolor+ink+gold)

▼ Happy New Year (watercolor+ink+gold)

"不同的文化，不同的習俗，有著同樣的歡樂。"

▲ Lion and Tiger Dance Show (watercolor+ink+gold)

Line Art

線條藝術 以簡潔的線條表達形式，突出輪廓和結構。簡約而直接的視覺效果，讓觀者可以自由地填入想像和感受。

Songkran Festival

Line art, illustration for children, children in colourful clothing play in water, in the style of dansaekhwa, Songkran festival, Cultural Illustration

Tomb-sweeping Day

April Fools' Day

Mother's Day

International Women's Day

▼ Children's Day

"溫馨與愛，是人類共同的語言。"

▲ A Christmas Party

Day of the Dead

Halloween

▼ Holi

"節慶的教育價值，在於引導人們把生活中的未知與恐懼，
轉化為歡樂與趣味。"

▲ Albuquerque International Balloon Fiesta

Picture book illustration, children in a village, watercolor painting, in the style of light orange and light brown, pattern-based painting, dolly kei, festive atmosphere, cultural documentation, group material, children's book illustrations

Storybook Magic: Unleashing Childhood Wonder through AI Illustrations

The Mythical Creature

The Fluffy Family

Enchanted Child

Cartoonish Illustration

卡通插畫 是繪本常見風格之一，以鮮豔的色彩和生動的形象深受喜愛。動物擬人化和表情誇張化經常用於創建氣氛和傳達情緒。有時，藝術家會使用不自然的色彩來強調某些元素，以幫助讀者理解，使之成為一種富有吸引力的視覺藝術手法。

Seasons & Life

Cartoonish Illustration, children drawing set depicting four seasons and different stages of life span from birth to death, in the style of vibrant airy scenes, squiggly line style, organic material, free brushwork, folkloric themes, shaped canvas

Japanese Speed Train

Threesome Camp

Mighty Mice Band

▼ Wondrous Tree House

"充滿趣味與超越現實的故事，帶給我們歡樂與
美好的回憶。"

▲ Fitness Run

Realistic Illustration

寫實插畫 以高度的真實再現和細緻的明暗表現吸引人們。此風格致力於真實比例和色彩的精確呈現，並經常與其他風格結合，創造出獨特的風味。它尤其適用於較大的兒童讀物，不僅支持文字理解，也能吸引和保持讀者的注意力。

The Red Dress Gal

Close up, realistic illustration of a chinese ghost myth, little girl with Oshiroi makeup, red lips, pale face, Phil Couture, naturalism, realism, dressed in red dress standing on a deserted mountain track, line art style, black lines with white background

Anticipation Lies

Grandma's Place

Young Geisha

▼ Baby Girl Reading Book

"寫實的人與物，逼真地呈現我們的內心世界，彷彿把
夢境帶到現實中。"

▲ Thoughts Running

Line Drawing Illustration

線描繪畫 / 線描繪圖 以簡單線條表達形狀與結構，強調形狀與空間關係，創造出強烈視覺效果。常見於塗色書，以簡潔的線條和色塊強化主題概念。此風格透過線條的俐落和簡單，能有效傳達出想法。

Ecological Water Plants

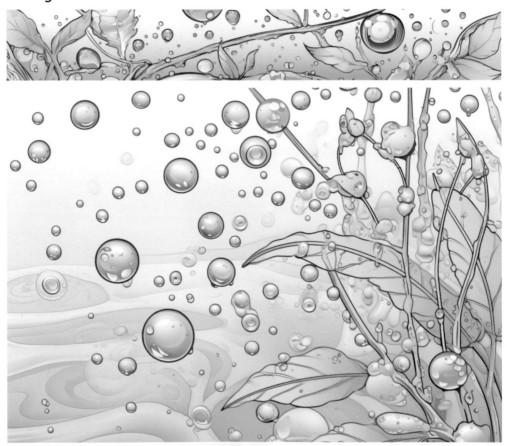

Line drawing illustration, kidz doodle drawing book screenshot thumbnail, in the style of naoko takeuchi, sky-blue and green, ecological art, water drops, pencil sketch, eye-catching composition, poetic still lifes

Kawaii Animals

Bird on Tree Branch

Street Scene

▼ A Flying Witch

"簡單且輕鬆的構圖，帶來滿滿的奇幻與想像空間。"

▲ Extended Playhouse

Abstract Illustrations

抽象插畫 著重在形狀、色彩和線條的組合，強化主題和情緒的傳達。它不追求實物描繪，而是以極簡和誇張的視覺元素構成，讓觀者從中得到自由的解讀。這種風格雖難以定義，但一看即可認出其獨特性。

The Zodiac Dog

Abstract illustration, chinese zodiac animals, colorful storytelling, animalpunk, lively storytelling, arts and crafts movement

Oh My Cats

Picnic Under Sakura

The Night Guards

▼ Under the Sea

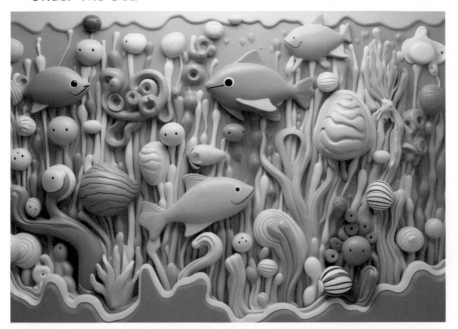

"超出現實的風格，不只帶來視覺的震撼，更驅動人類的
好奇心。"

▲ The Peeping Eyes

Vintage Illustration

復古插畫 擅長融合過去與現在，低調而引人入勝。線條細膩，色調淡化，彷彿引領我們回到更簡單的時代。不同時期的復古風格有其獨特的韻味，利用鉛筆繪製，深受讀者喜愛，喚起對過去的回憶與現今的理解。

Humpty Dumpty

Vintage Illustration, Humpty Dumpty wearing a suit and tall black hat sits on a wall, Humpty Dumpty had a great fall, overwhelming style, Illustration Mignonne

The Wizard of Oz

The Angels

Charlotte's Web

"精緻且寫實的插畫，讓這些人物栩栩如生；令人有進入
故事的時空，身歷情境的感覺。"

▲ Three Little Pigs

Picture Book Illustration

繪本插畫 巧妙結合多種藝術元素，如鉛筆筆觸、水彩與粉彩，以暖色調和童趣呈現故事。其視覺敘事引發讀者想像，深度與細節環環相扣，擴大視野，影響情感思緒。

Cats on Boat

Picture book illustration, painting shows a boat under the moon lit up with stars, in the style of playful and whimsical depictions of animals, villagecore, miwa komatsu, panorama, cute and dreamy, celebration of rural life, gabriel bá

BBQ During Moon Cake Festival

Nursery Animals

Tetyana Bear in Rain

▼ The Little Polka Dot Bird

"滿滿的歡樂，滿滿的溫馨與趣味；這是伴隨很多人
成長的畫風。"

▲ The Playful Elephant

People, prosperity, planet, peace, partnership, showing sustainable development goals, panoramic view, dreamy light, shutter speed 1/125, F/2.4, iso 200, Nikon, blurred background, super details

Sustainable Visions: The Importance and Promotion of SDGs in AI Art

Save World Water

Conservation

Ecological Recycle

Houses on Dirt

Wars Collapsed Walls

▼ United Colors

"這是我們所在的地球村，充滿了形形色色的人與事；即使
有許多的不幸，依然能夠看到曙光與美好的色彩。"

▲ Drought

Photobash Style

Photobash 風格 是將照片 (Photo) 做組合 (bash)，透過 AI 技術，更熟練地將指定場景拼湊並繪製成一幅寫實且細膩的插畫。此風格獨特在於其高度的真實性和細節豐富，尤其適合寫實主題的示意圖。*Photobash* 風格不僅是對照片元素的巧妙組合，也是對現實與想像的創新融合，提供了一種超越真實的視覺體驗。

#1 Eliminate Poverty

Photobash style of a young skinny child in skimpy clothing sitting in dusty furniture in kenya, in the style of domestic intimacy, tanbi kei, domestic interiors, leather/hide, associated press photo, happenings

#2 Erase Hunger

#3 Establish Good Health & Well-Being

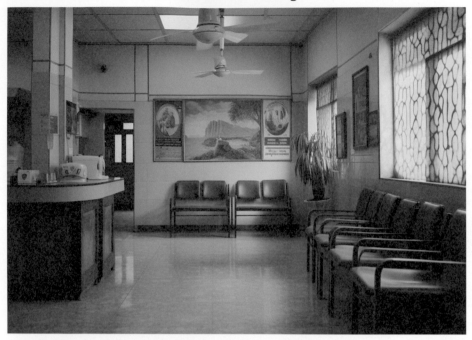

#4 Provide Quality Education

#5 Enforce Gender Equality

"我們生活在一個美的世界，人人都有機會；只要我們的
心是美的，世界就會更美。"

#6 Improve Clean Water & Sanitation

Bold & Bright Style

大膽鮮豔風格 以鮮明色調和強烈對比引人注目，為作品帶來強烈視覺與情感效果。其大膽用色和混搭風格挑戰視覺舒適區，營造出充滿活力和深度的畫面。該風格意味的是要走出我們的舒適區，混搭出新的風格。大膽的用色，讓畫面深度跳出來。

#7 Grow Affordable & Clean Energy

Bold and bright style, clean, renewable energy technologies including wind, solar, hydro, geothermal, bioenergy & nuclear

#9 Increase Industry, Innovation & Infrastructure

#10 Reduce Inequality

#11 Mobilize Sustainable Cities & Communities

"世界真奇妙：由於持續的努力與創新，這個世界充滿了
驚喜與神奇；把得來不易的幸福傳遞給需要的人，是我們
應該努力的下一步。"

#12 Influence Responsible Consumption & Production

Pointillism

點彩畫派/點描派風格 以其獨特的技法和視覺效果而受到讚揚。點彩適合描繪戶外風景、肖像和海景。在表面塗上小筆或小點的顏色,使其在視覺上從遠處看似融合在一起。

#13 Organize Climate Action

Pointillism, integrate climate change measures, Climate Action and Synergies

#14 Develop Life Below Water

#15 Advance Life on Land

#16 Guarantee Peace, Justice & Strong Institutions

#17 Build Partnerships for the Goals

White background, simple colors, in the library, children are happily reading books, children's style crayon doodle drawings, --no watermark

Children Drawing Style, iPhone UI/UX Design

Line Art, iPhone UI/UX Design

iPhone 14 pro fashion case, Fashion illustration, a cute cat, in the style of digital illustration, victor Mizintsev, watercolor art

Dashboards smart libraries for phone simulations book circulation application design, modern, dimensions

Dashboard smart home app design in telephone mockups, modern, dimensional

Stained Glass Style

彩繪玻璃風格 以色彩繽紛和複雜幾何圖案
著稱。透過組合不同色彩和玻璃片，創造出華
麗、獨特的藝術效果。常見於教堂、窗戶和藝
術品中。

Stained Glass Style Butterfly

Stained glass style butterfly

Frame your Creations & Place them in a Room

1. 放大您想放在相框內的圖像

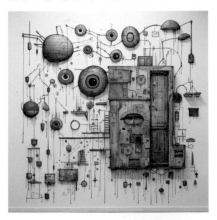

/imagine prompt Simple children's drawing guide for A conceptual artwork using found objects and unconventional materials to create a thought-provoking installation

2. 選擇 🔍 自定義縮放選項 (Custom Zoom option)

3. 刪除您第一次使用的提示 (prompt)，並用以下文字代替："A framed picture on the wall"

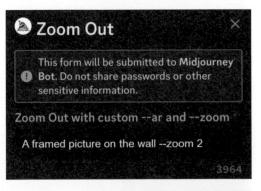

X-ray Double Exposure

X-ray 雙重曝光 結合 X 光影像和其他圖像元素，以創造出獨特的視覺效果。透過 AI art 技術的應用，使不同圖像疊加，呈現出透視效果和神秘感。*X-ray* 雙重曝光作品讓觀者感受到探索隱藏層面和多重意義的體驗。

Apples

Guitar

X-ray double exposure of a Guitar

X-ray double exposure of NYC, neon art light, volumetric lighting, hyper detailed

X-ray style, Stained Glass Style Butterflies

Washi Tape Style

和紙膠帶風格 以日本和紙藝術元素為特點，創造出親切、手工和可愛的氛圍。常見於手帳、手工藝品和設計，為作品增添獨特魅力。

Birds Sitting on a Fence in a Meadow

Birds sitting on a fence in a meadow, in the style of washi tape, bright colours and patterns, intricate details

A girl and a seated cat, in the style of washi tape, bright colours and patterns

Seaside village, in the style of washi tape, bright colours and patterns, intricate details

Super Cute Animal Stickers

貼紙設計 以活潑的色彩和簡潔的圖像傳達直接的訊息。強調獨特性和創新性，色彩、形狀和圖案的巧妙組合吸引觀者目光。它不僅視覺上吸引人，更是創意和訊息的有效傳遞方式。

Cute Baby Birds

Cute Baby Sheep (Thumbprint Art)

Sticker, a super cute baby sheep, vector, thumbprint art

Sticker Design, Lined Paper for Writing

"散發歡樂及正能量的簡單創意，是現代社會最需要的
光與熱。"

Sticker Design

貼紙設計 以活潑的色彩和簡潔的圖像傳達直接的訊息。強調獨特性和創新性，色彩、形狀和圖案的巧妙組合吸引觀者目光。它不僅視覺上吸引人，更是創意和訊息的有效傳遞方式。

Design a Cute Hacker(s) Sticker

Design a sticker of cute hacker with black hair and eyeglasses coding using a cool keyboard, the developer sit on a sofa, die cut sticker, white background

Design a Set of Stickers Inspired by Vintage Halloween Decorations

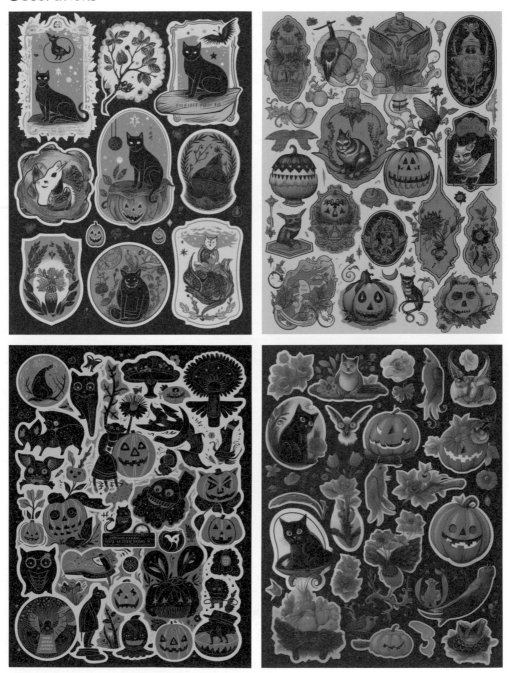

Signet Rings

Silver "LOVE" signet ring

A ring of rose gold and glass rings make the most stunning jewelry, gorgeous thick opals sparkle ethereal sparkles complex translucent opals intricate, crazy details, glazing, inner glow, strobe lights

T-Shirt Graphics

T-Shirt Graphics 以其獨特的平面設計和視
覺吸引力受到廣泛喜愛。這種風格以鮮明的圖像和
創意的構圖為特點，將 T 恤變成了一個行走的藝術
展示。從簡單的圖案到複雜的插圖，T 恤圖形風格
能夠展現個人風格和表達獨特的訊息。

T-shirt

Tshirt vector, white roses graphic, vivid colors, detailed

Cute Tiny Isometric Architecture

超萌微型等角建築風格 以可愛且精緻的尺度，展現了建築物的立體之美。色彩鮮明、細節生動的小型建築呈現，營造出歡樂與趣味橫生的視覺饗宴。

A 3D Isometric Aqua and White Miniature Mosque

A Cute 3D Isometric View of Library

Create a cute 3d isometric view of library with soft colors and materials

A submarine restaurant, Miyazaki Style, isometric view

Isometric clean art of a modern tiny library designed by kengo kuma, blender

Isometric Cutaway

等角切面圖風格 以一種獨特的視角展
示結構的內部細節。此風格捕捉了物體內
部的精密結構，對觀者揭示了看不見的細
節和功能。

"繪畫帶來無限的想像空間，無限的可能性與創作機會；唯
一的限制是你的想像力。"

The Inside of a Watermelon

Isometric clean pixel art image profile of the inside of a watermelon

Toy Cars (Cartoon Style)

Clipart vibrant and colorful, clean, tiny cute cars for kids on a white background, Scandinavian style, minimalist, shading, high detail, high quality, high resolution, cartoon style

Paper Art

Paper Art 被稱為 **紙藝** 或 **紙雕藝術**，為一種以紙張作為主要媒介的藝術形式，包括剪紙、折紙、繪畫、拼貼、紙塑等各種表現形式。紙藝的製作技巧和風格可能因文化、工藝和藝術家的創作意圖而變化。從簡單的手工藝到複雜的藝術裝置，紙藝都展現了其獨特的創新性和多樣性。其包含 paper cut art, origami, paper crafts, quilted paper art, paper sculpture, collage 等。

/imagine prompt <any art style> , a cute cat/ cute cats in the garden

Paper Art

Paper Cut Art

Quilling Paper Art

Papercut Sculpture

Paper Crafts

Collage of cute a {happy mouse, majesty tiger} made from tiny pieces of paper with numbers, wide angle, full shot

Bicycle

Garden

Santorini, Greece

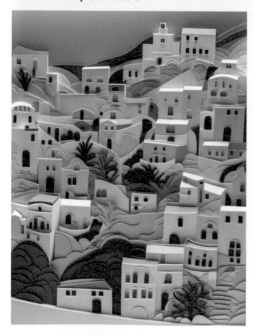

Cappadocia, Turkey

ASCII Art

ASCII 藝術 以拉丁字母、數字、標點符號以及其他特殊符號等字元被巧妙地組合是使用電腦上標準的字元編碼方式來創作藝術作品,以此建立吸引人的視覺化圖像或設計。此種藝術形式既能簡單如表情符號,也能複雜到形成具象或抽象的圖形,表現出極高的創作靈活性。

Moon

Girl

URL ASCII art drawing style, anime, asian girl, black and white, asian face, in the style of text-based mixed media, martin rak, chie yoshii, computer art, eye-catching composition, translucent sparkle resin, metal base, flickr, adrian donoghue

ASCII style, pop art, wallpaper, stained glass moon

Dreamy Watercolor Scenes

夢幻水彩場景 以其溫柔、流動的色彩營造出令人著迷的氛圍。這種風格利用水彩的獨特性質，呈現一種夢幻般的視覺效果，將觀者帶入一個寧靜而美麗的世界。

Boats in the Harbor

Children Playing at the Beach

Light watercolor of children playing on the beach, white background, few details, ink+gold, colorful, dreamy Studio Ghibli

Light watercolor, a couple of lovers, at the beach, few details, dreamy

Fixed camera, watercolor sketch, simple Clip art, an elegant woman in an elegant dress on the streets of Paris, 16k, 70s, elegant attire, femme fatale

Character Concept Sheet

角色概念表 呈現角色的外觀、個性和背景等細節。這種詳盡的設計提供全面的角色描述，幫助藝術家和創作者清晰呈現角色的視覺和故事元素。

Robot Character Design

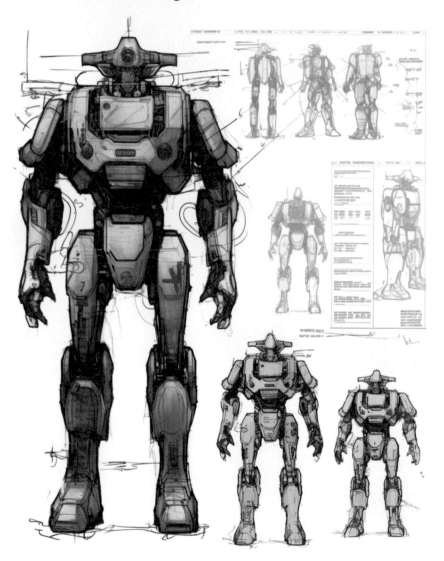

Robot character design, concept design sheet, white background

Female robot close up character design, multiple concept designs, concept design sheet, white background, style of Yoshitaka Amano

Net Art

網路藝術 能展現出一種難以捉摸的特質。透過網路素材和網站數據的巧妙運用,展現其獨特的動態性和瞬息萬變的特性。網路藝術不僅反映了當代科技的影響,更突顯藝術在虛擬空間中的無限延伸,打破傳統藝術的界限。

Computing & Machine Aesthetics

Utopia of Internet

Net art of utopia, internet and coding, ethereal background

Mathematical Structures

數學結構 對於數字、空間或邏輯關係的相互聯繫至關重要。但，目前尚不清楚 AI 對這一概念的理解程度如何。我們可以透過多方面的嘗試並利用墊圖的方式，進行探索更多有趣的想法和創新的發想。

Chaotic Atom Structure

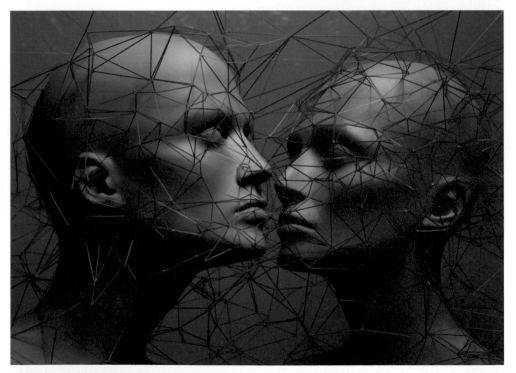

Mathematical Structures of an image of a paramid, with lines connected to from all angles, in the style of dark azure and orange, precise, daz3d, constructivist roots, serene faces, net art, arthur tress

Voxel Art

體素藝術 適合用於探索低分辨率或複古外觀的藝術作品，如 8 位或 16 位調色板的像素藝術、現代像素與體素藝術。此風格在視覺化、遊戲設計、動畫和其他多媒體應用方面受到歡迎，提供了一種獨特且引人入勝的視覺體驗。

Farm Animals

Stairs to Unknown

Voxel art of a maze of stairs made in colorful tiles and blue lights, in the style of light green and pink, rendered in maya, paper sculptures, animated gifs, surreal city scenes, jazzy interiors, zigzags

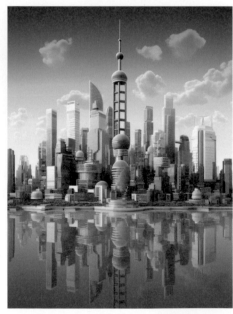

Voxel art of shanghai skyline buildings, jazzy interiors, zigzags

Sand Art

沙畫 是將沙塑造成藝術形式的實踐，例如刷沙、沙雕、沙畫或製作沙瓶。可用於微型建築、人物等創作。

Eritrean Sand Flowers

The Thinker

Sand art, museum masterpiece of a figurine sculpture

3D Rendering

3D 彩現 /3D 渲染 是一個跨界的視覺技術，廣泛應用於電腦與電動遊戲、類比、電影、電視特效和視覺化設計等。透過此技術能將想像力轉化為逼真的視覺效果，增強遊戲畫面中的真實感，也能提升電影和電視的震撼場景，展現了創意和表現力。

Blockchain of LOVE

Crescent Sun Moon Design

3d rendering image of an astrological arrangement of gold and silver in the form of taiji theme, depicting sun and moon crest, on white background

Cryptidcore

隱核風格 於 2014 年由 tumblr 使用者間所流行的一種美學。其通常會有一種複古的氛圍，涉及神話、民間和人類未解之謎，主題大多與都市傳說有關。該風格與 Cottagecore 和 Weirdcore 之間也有點相關。可以嘗試沿著這些風格進行探索，深入挖掘我們腦海中那些奇妙與神秘的超自然現象。

Meta Coin

Cryptidcore, metallic silicon coin made of code with carbonite texture

The Knowing Binary Eye

Cryptidcore, the all knowing binary eye with blue lines of binary numbers, in the style of organic material, detailed backgrounds, argus c3, technological marvels, light black and gold, psychological phenomena illustrations, rim light

Alcohol Ink

酒精墨水畫 是一種自由而富有創意的藝術表達方式，它讓每一位創作者，不論繪畫技巧如何，都能展現出獨特的美感。透過酒精的流動和風力的引導，墨水在畫布上擴散開來，形成大理石紋、年輪流線、放射線和渲染等效果。當我們透過 Midjourney 施展該風格創意時，不妨可以嘗試添加指令如 --chaos 或 --stylize 來獲得更多的創意。

Stunning Flowers

Wabi-sabi Philosophy

Marble Floral Spread

Alcohol Ink, Japanese philosophy of wabi-sabi

Alcohol Ink, marble white, dark grey, black and gold

Alcohol Ink, 4 cliparts, watercolor wedding magic clipart, cute set bundle, simple drawing, line art, white background, cartoon --v 4

Alcohol Ink, four images showing women in different dresses walking between shadows, spectacular backdrops, mystical portraits, ottoman art, fantastical compositions --v 4

Niji · journey Original style

原始風格 將多種藝術風格的特點融合在一起，能保留了它們的精髓，同時進行優化和整合。該風格不僅能創造出一個多元而和諧的視覺效果，它還尊重原始的精神。

Dark, dramatic contrast, red hair {girl, man} girl, an expansive view of, the low-purity tone, sun light, 16K, best quality --q 5 --s 100 --niji 5 --style original

Niji · journey Scenic style

場景風格 是一種強調畫面完整性和故事性的藝術表現手法。該風格的特點能精細地繪製畫面背景，讓整個畫面充滿生氣，富有電影感。對於描繪複雜場景的作品而言，這種風格尤為合適。場景風格擅長將環境與人物完美融合，創造出和諧的視覺效果。

Lake under the stars, rococo, ultrawide shot --q 5 --s 100 --niji 5 --style scenic

Niji・journey Default style

默認風格 通常是指一種基本、標準或常見的繪畫或藝術風格；其是大眾熟悉且被廣泛使用的風格。

A beautiful girl standing on the clouds, busts, gorgeous white dress, volume trice lighting, cold light, back light, symmetrical the composition, trending on artstaion, 16k, best quality --q 5 --s 100 --niji 5 --Default style

A beautiful girl is resting in the court, gorgeous dress,volumetrice lighting, gloden hour light, epic detail, trending on artstaion, 4k, best quality --q 5 --s 100 --niji 5

Niji · journey Expressive style

風格展現力 不只是一種藝術表現方式，也是一種深入角色的內心世界和情感的探索方式。該風格能對客觀事物的精準描繪，也能透露藝術家的內心情感和對角色真實內心的理解。透過捕捉和呈現深層的情感，它將藝術提升到了一個新的層次，使之不只是觀看的對象，而是一種能觸動人類內心的直接表達方式。

Holy Angel

Albinism girl, holy angel, white wings, beautiful red-eyed, long white hair, under the moon, white flowers on the floor with some water, gorgeous dress, gold white silver, rococo, ethereal, close up straight to the face portrait, hands-near-the-face portrait, quiet, 8k, high detail, best quality --q 5 --s 500 --niji 5 --style expressive

Delicate beautiful quiet girl like a doll sit down on the high place, holy dark angel with black wings, beautiful red-eyed, long black hair, under the moon, red flowers on the floor with some water, gorgeous black dress, intricate details, gold white silver, Rococo,ethereal,close up straight to the face portrait,hands-near-the-face portrait,8k,high detail,best quality --q 5 --s 100 --niji 5 --style expressive

Niji・journey Cute style

可愛風格 是一種以迷人的可愛、親和力和童趣為主要特點的藝術風格；透過將主題、角色或物品描繪得具有吸引力。此風格往往能創造出輕鬆、愉快的氛圍，其也能引發人們內心的童貞和喜愛。

A Beautiful Mermaid in the Sea

A beautiful mermaid in the sea, pink and blue --q 5 --s 100 --niji 5 --style cute

A cute chibi girl in the sea, blue and purple --q 5 --s 100 --niji 5 --style cute

A cute girl in the forest, pink and purple --q 5 --s 100 --niji 5 --style cute

Expressive Comic Panels

The young girl reading a book after finishing her breakfast, in the style of anime aesthetic, jarring juxtapositions, i can't believe how beautiful this is, expressive comic panels, light navy and light yellow, realistic yet romantic, group f/64 --niji 5

A page of a girl and a dog in the garden, with black dog, in the style of romantic manga, garden portraits, i can't believe how beautiful this is, shiny eyes, panoramic scale, garden views, official art --niji 5

國家圖書館出版品預行編目（CIP）資料

未來藝世界 : AI 繪圖新旅程 = Arts Styles of the Future:
A Non-Technical Introduction to AI Drawing / 黃國禎,
涂芸芳, 金玲, 楊梅伶, 白璃, 邱敏棋合著 . -- 初版 . --
新北市 : 斑馬線出版社, 2023.09
　　面；　公分

　ISBN 978-626-96854-8-6（平裝）

　1.CST: 數位藝術 2.CST: 人工智慧 3.CST: 數位學習

312.86　　　　　　　　　　　　112012597

未來藝世界：AI 繪圖新旅程
Arts Styles of the Future: A Non-Technical Introduction to AI Drawing

作　　者：黃國禎、涂芸芳、金　玲、楊梅伶、白　璃、邱敏棋
總 編 輯：施榮華
封面設計：余佩蓁

發 行 人：張仰賢
社　　長：許　赫
副 社 長：龍　青
出 版 者：斑馬線文庫有限公司
法律顧問：林仟雯律師

斑馬線文庫
通訊地址：234 新北市永和區民光街 20 巷 7 號 1 樓
連絡電話：0922542983

製版印刷：龍虎電腦排版股份有限公司
出版日期：2023 年 9 月
I S B N：978-626-96854-8-6
定　　價：460 元